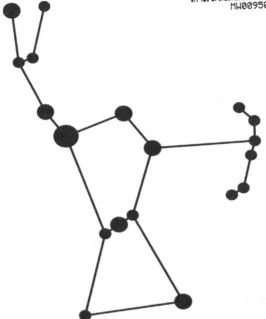

This Astronomy Observation Log Book Belongs To:

Name:_____

Phone:_____

Email:_____

Astronomy Observation Sheet

Name:		Date: / /	Time: :
Scope:		Seeing:	
Site & Notes:			

Notes:

NGC		Filter	
FOV		MAG	

Notes:

NGC		Filter	
FOV		MAG	

Notes:

NGC		Filter	
FOV		MAG	

Astronomy Observation Log

Name:		Location:		Page:
Index	**Details**		**Notes**	
Object	Date: / / Time : Eye Piece: Seeing: Instrument Type:			
Object	Date: / / Time : Eye Piece: Seeing: Instrument Type:			
Object	Date: / / Time : Eye Piece: Seeing: Instrument Type:			
Object	Date: / / Time : Eye Piece: Seeing: Instrument Type:			
Object	Date: / / Time : Eye Piece: Seeing: Instrument Type:			
Object	Date: / / Time : Eye Piece: Seeing: Instrument Type:			
Object	Date: / / Time : Eye Piece: Seeing: Instrument Type:			
Object	Date: / / Time : Eye Piece: Seeing: Instrument Type:			
Object	Date: / / Time : Eye Piece: Seeing: Instrument Type:			
Object	Date: / / Time : Eye Piece: Seeing: Instrument Type:			

Observation Log & Sketch Template

Observer:_____ Date:_____ / / _____ Time:_____:_____

Location:_____ Seeing (1 – 10): _____ Transparency (1 – 5)_____

Object:_____ Constellation:_____

R.A_____hrs_____min Dec_____deg_____min Magnitude:_____Size:_____

Observing Equipment :_____ Eye Pieces:_____

Filters :_____ Images:_____

Field Drawing

Low Power Ocular　　　　**High Power Ocular**

Description & Notes

_____
_____
_____
_____
_____
_____
_____
_____

Astronomy Observation Sheet

Name:	Date: / /	Time: :
Scope:	Seeing:	
Site & Notes:		

Notes:

NGC	Filter
FOV	MAG

Notes:

NGC	Filter
FOV	MAG

Notes:

NGC	Filter
FOV	MAG

Astronomy Observation Log

Name:		Location:	Page:

Index	Details	Notes
Object	Date: / / Time : Eye Piece: Seeing: Instrument Type:	
Object	Date: / / Time : Eye Piece: Seeing: Instrument Type:	
Object	Date: / / Time : Eye Piece: Seeing: Instrument Type:	
Object	Date: / / Time : Eye Piece: Seeing: Instrument Type:	
Object	Date: / / Time : Eye Piece: Seeing: Instrument Type:	
Object	Date: / / Time : Eye Piece: Seeing: Instrument Type:	
Object	Date: / / Time : Eye Piece: Seeing: Instrument Type:	
Object	Date: / / Time : Eye Piece: Seeing: Instrument Type:	
Object	Date: / / Time : Eye Piece: Seeing: Instrument Type:	
Object	Date: / / Time : Eye Piece: Seeing: Instrument Type:	

Observation Log & Sketch Template

Observer:_____ Date:____/____/_____ Time:____:_____

Location:_____ Seeing (1 – 10):_____ Transparency (1 – 5)_____

Object:_____ Constellation:_____

R.A_____hrs_____min Dec_____deg_____min Magnitude:_____Size:_____

Observing Equipment :_____ Eye Pieces:_____

Filters :_____ Images:_____

Field Drawing

Low Power Ocular   High Power Ocular

Description & Notes

_____
_____
_____
_____
_____
_____
_____

Astronomy Observation Sheet

Name:	Date: / /	Time: :
Scope:	Seeing:	
Site & Notes:		

Notes:

NGC	Filter
FOV	MAG

Notes:

NGC	Filter
FOV	MAG

Notes:

NGC	Filter
FOV	MAG

Astronomy Observation Log

ame:	Location:	Page:

Index	Details	Notes
Object	Date: / / Time : Eye Piece: Seeing: Instrument Type:	
Object	Date: / / Time : Eye Piece: Seeing: Instrument Type:	
Object	Date: / / Time : Eye Piece: Seeing: Instrument Type:	
Object	Date: / / Time : Eye Piece: Seeing: Instrument Type:	
Object	Date: / / Time : Eye Piece: Seeing: Instrument Type:	
Object	Date: / / Time : Eye Piece: Seeing: Instrument Type:	
Object	Date: / / Time : Eye Piece: Seeing: Instrument Type:	
Object	Date: / / Time : Eye Piece: Seeing: Instrument Type:	
Object	Date: / / Time : Eye Piece: Seeing: Instrument Type:	
Object	Date: / / Time : Eye Piece: Seeing: Instrument Type:	

Observation Log & Sketch Template

Observer:_____ Date:____ / ____ / _____ Time:____:_____

Location:_____ Seeing (1 – 10): _____ Transparency (1 – 5)_____

Object:_____ Constellation:_____

R.A_____ hrs_____min Dec_____deg_____min Magnitude:_____Size:_____

Observing Equipment :_____ Eye Pieces:_____

Filters :_____ Images:_____

Field Drawing

Low Power Ocular   **High Power Ocular**

Description & Notes

_____
_____
_____
_____
_____
_____
_____
_____

Astronomy Observation Sheet

Name:	Date: / /	Time: :
Scope:	Seeing:	
Site & Notes:		

Notes:

NGC	Filter
FOV	MAG

Notes:

NGC	Filter
FOV	MAG

Notes:

NGC	Filter
FOV	MAG

Astronomy Observation Log

Name:		Location:	Page:

Index	Details	Notes
Object	Date: / / Time : Eye Piece: Seeing: Instrument Type:	
Object	Date: / / Time : Eye Piece: Seeing: Instrument Type:	
Object	Date: / / Time : Eye Piece: Seeing: Instrument Type:	
Object	Date: / / Time : Eye Piece: Seeing: Instrument Type:	
Object	Date: / / Time : Eye Piece: Seeing: Instrument Type:	
Object	Date: / / Time : Eye Piece: Seeing: Instrument Type:	
Object	Date: / / Time : Eye Piece: Seeing: Instrument Type:	
Object	Date: / / Time : Eye Piece: Seeing: Instrument Type:	
Object	Date: / / Time : Eye Piece: Seeing: Instrument Type:	
Object	Date: / / Time : Eye Piece: Seeing: Instrument Type:	

Observation Log & Sketch Template

Observer:_____ Date:____/___/_____ Time:____:_____

Location:_____ Seeing (1 – 10):_____ Transparency (1 – 5)_____

Object:_____ Constellation:_____

R.A_____hrs_____min Dec_____deg_____min Magnitude:_____Size:_____

Observing Equipment :_____ Eye Pieces:_____

Filters :_____ Images:_____

Field Drawing

Low Power Ocular   **High Power Ocular**

Description & Notes

_____
_____
_____
_____
_____
_____
_____

Astronomy Observation Sheet

Name:	Date: / /	Time: :
Scope:	Seeing:	
Site & Notes:		

Notes:

NGC	Filter
FOV	MAG

Notes:

NGC	Filter
FOV	MAG

Notes:

NGC	Filter
FOV	MAG

Astronomy Observation Log

Name:	Location:	Page:

Index	Details	Notes
Object	Date: / / Time : Eye Piece: Seeing: Instrument Type:	
Object	Date: / / Time : Eye Piece: Seeing: Instrument Type:	
Object	Date: / / Time : Eye Piece: Seeing: Instrument Type:	
Object	Date: / / Time : Eye Piece: Seeing: Instrument Type:	
Object	Date: / / Time : Eye Piece: Seeing: Instrument Type:	
Object	Date: / / Time : Eye Piece: Seeing: Instrument Type:	
Object	Date: / / Time : Eye Piece: Seeing: Instrument Type:	
Object	Date: / / Time : Eye Piece: Seeing: Instrument Type:	
Object	Date: / / Time : Eye Piece: Seeing: Instrument Type:	
Object	Date: / / Time : Eye Piece: Seeing: Instrument Type:	

Observation Log & Sketch Template

Observer:_____ Date:_____ / / _____ Time:_____:_____

Location:_____ Seeing (1 – 10): _____ Transparency (1 – 5)_____

Object:_____ Constellation:_____

R.A _____ hrs _____ min Dec _____ deg _____ min Magnitude:_____ Size:_____

Observing Equipment :_____ Eye Pieces:_____

Filters :_____ Images:_____

Field Drawing

Low Power Ocular   **High Power Ocular**

Description & Notes

_____
_____
_____
_____
_____
_____
_____
_____

Astronomy Observation Sheet

Name:	Date: / /	Time: :
Scope:	Seeing:	
Site & Notes:		

Notes:

NGC	Filter
FOV	MAG

Notes:

NGC	Filter
FOV	MAG

Notes:

NGC	Filter
FOV	MAG

Astronomy Observation Log

ame:	Location:	Page:

Index	Details	Notes
Object	Date: / / Time : Eye Piece: Seeing: Instrument Type:	
Object	Date: / / Time : Eye Piece: Seeing: Instrument Type:	
Object	Date: / / Time : Eye Piece: Seeing: Instrument Type:	
Object	Date: / / Time : Eye Piece: Seeing: Instrument Type:	
Object	Date: / / Time : Eye Piece: Seeing: Instrument Type:	
Object	Date: / / Time : Eye Piece: Seeing: Instrument Type:	
Object	Date: / / Time : Eye Piece: Seeing: Instrument Type:	
Object	Date: / / Time : Eye Piece: Seeing: Instrument Type:	
Object	Date: / / Time : Eye Piece: Seeing: Instrument Type:	
Object	Date: / / Time : Eye Piece: Seeing: Instrument Type:	

Observation Log & Sketch Template

Observer:_____ Date:____/____/_____ Time:____:____

Location:_____ Seeing (1 – 10): _____ Transparency (1 – 5)_____

Object:_____ Constellation:_____

R.A_____hrs_____min Dec_____deg_____min **Magnitude:**_____Size:_____

Observing Equipment :_____ Eye Pieces:_____

Filters :_____ Images:_____

Field Drawing

Low Power Ocular   **High Power Ocular**

Description & Notes

_____
_____
_____
_____
_____
_____
_____
_____

Astronomy Observation Sheet

Name:	Date: / /	Time: :
Scope:	Seeing:	
Site & Notes:		

Notes:

NGC	Filter
FOV	MAG

Notes:

NGC	Filter
FOV	MAG

Notes:

NGC	Filter
FOV	MAG

Astronomy Observation Log

Name:		Location:		Page:
Index	**Details**		**Notes**	
Object	Date: / / Time : Eye Piece: Seeing: Instrument Type:			
Object	Date: / / Time : Eye Piece: Seeing: Instrument Type:			
Object	Date: / / Time : Eye Piece: Seeing: Instrument Type:			
Object	Date: / / Time : Eye Piece: Seeing: Instrument Type:			
Object	Date: / / Time : Eye Piece: Seeing: Instrument Type:			
Object	Date: / / Time : Eye Piece: Seeing: Instrument Type:			
Object	Date: / / Time : Eye Piece: Seeing: Instrument Type:			
Object	Date: / / Time : Eye Piece: Seeing: Instrument Type:			
Object	Date: / / Time : Eye Piece: Seeing: Instrument Type:			
Object	Date: / / Time : Eye Piece: Seeing: Instrument Type:			

Observation Log & Sketch Template

Observer:_____ Date:____/__/_____ Time:____:_____

Location:_____ Seeing (1 – 10):_____ Transparency (1 – 5)_____

Object:_____ Constellation:_____

R.A_____hrs_____min Dec_____deg_____min Magnitude:_____Size:_____

Observing Equipment :_____ Eye Pieces:_____

Filters :_____ Images:_____

Field Drawing

Low Power Ocular　　　　　　　　　　**High Power Ocular**

Description & Notes

_____

_____

_____

_____

_____

_____

_____

_____

Astronomy Observation Sheet

Name:	Date: / /	Time: :
Scope:	Seeing:	
Site & Notes:		

Notes:

NGC	Filter
FOV	MAG

Notes:

NGC	Filter
FOV	MAG

Notes:

NGC	Filter
FOV	MAG

Astronomy Observation Log

Name:	Location:	Page:

Index	Details	Notes
Object	Date: / / Time : Eye Piece: Seeing: Instrument Type:	
Object	Date: / / Time : Eye Piece: Seeing: Instrument Type:	
Object	Date: / / Time : Eye Piece: Seeing: Instrument Type:	
Object	Date: / / Time : Eye Piece: Seeing: Instrument Type:	
Object	Date: / / Time : Eye Piece: Seeing: Instrument Type:	
Object	Date: / / Time : Eye Piece: Seeing: Instrument Type:	
Object	Date: / / Time : Eye Piece: Seeing: Instrument Type:	
Object	Date: / / Time : Eye Piece: Seeing: Instrument Type:	
Object	Date: / / Time : Eye Piece: Seeing: Instrument Type:	
Object	Date: / / Time : Eye Piece: Seeing: Instrument Type:	

Observation Log & Sketch Template

Observer:_____ Date:____/____/_____ Time:_____:_____

Location:_____ Seeing (1 – 10): _____ Transparency (1 – 5)_____

Object:_____ Constellation:_____

R.A_____hrs_____min Dec_____deg_____min Magnitude:_____Size:_____

Observing Equipment :_____ Eye Pieces:_____

Filters :_____ Images:_____

Field Drawing

Low Power Ocular   High Power Ocular

Description & Notes

Astronomy Observation Sheet

Name:	Date: / /	Time: :
Scope:	Seeing:	
Site & Notes:		

Notes:

NGC	Filter
FOV	MAG

Notes:

NGC	Filter
FOV	MAG

Notes:

NGC	Filter
FOV	MAG

Astronomy Observation Log

ame:	Location:	Page:

Index	Details	Notes
Object	Date: / / Time : Eye Piece: Seeing: Instrument Type:	
Object	Date: / / Time : Eye Piece: Seeing: Instrument Type:	
Object	Date: / / Time : Eye Piece: Seeing: Instrument Type:	
Object	Date: / / Time : Eye Piece: Seeing: Instrument Type:	
Object	Date: / / Time : Eye Piece: Seeing: Instrument Type:	
Object	Date: / / Time : Eye Piece: Seeing: Instrument Type:	
Object	Date: / / Time : Eye Piece: Seeing: Instrument Type:	
Object	Date: / / Time : Eye Piece: Seeing: Instrument Type:	
Object	Date: / / Time : Eye Piece: Seeing: Instrument Type:	
Object	Date: / / Time : Eye Piece: Seeing: Instrument Type:	

Observation Log & Sketch Template

Observer:_____   Date:____ / ____ / _____ Time:_____:_____

Location:_____   Seeing (1 – 10): _____ Transparency (1 – 5)_____

Object:_____   Constellation:_____

R.A_____ hrs_____min Dec_____deg_____min   Magnitude:_____Size:_____

Observing Equipment :_____   Eye Pieces:_____

Filters :_____   Images:_____

Field Drawing

Low Power Ocular   **High Power Ocular**

Description & Notes

_____
_____
_____
_____
_____
_____
_____
_____

Astronomy Observation Sheet

Name:	Date: / /	Time: :
Scope:	Seeing:	
Site & Notes:		

Notes:

NGC	Filter
FOV	MAG

Notes:

NGC	Filter
FOV	MAG

Notes:

NGC	Filter
FOV	MAG

Astronomy Observation Log

Name:	Location:	Page:

Index	Details	Notes
Object	Date: / / Time : Eye Piece: Seeing: Instrument Type:	
Object	Date: / / Time : Eye Piece: Seeing: Instrument Type:	
Object	Date: / / Time : Eye Piece: Seeing: Instrument Type:	
Object	Date: / / Time : Eye Piece: Seeing: Instrument Type:	
Object	Date: / / Time : Eye Piece: Seeing: Instrument Type:	
Object	Date: / / Time : Eye Piece: Seeing: Instrument Type:	
Object	Date: / / Time : Eye Piece: Seeing: Instrument Type:	
Object	Date: / / Time : Eye Piece: Seeing: Instrument Type:	
Object	Date: / / Time : Eye Piece: Seeing: Instrument Type:	
Object	Date: / / Time : Eye Piece: Seeing: Instrument Type:	

Observation Log & Sketch Template

Observer:_____ Date:____/____/____ Time:____:____

Location:_____ Seeing (1 – 10):____ Transparency (1 – 5)____

Object:_____ Constellation:_____

R.A____hrs____min Dec____deg____min **Magnitude:**_____Size:_____

Observing Equipment :_____ Eye Pieces:_____

Filters :_____ Images:_____

Field Drawing

Low Power Ocular **High Power Ocular**

Description & Notes

_____
_____
_____
_____
_____
_____
_____
_____

Astronomy Observation Sheet

Name:	Date: / /	Time: :
Scope:	Seeing:	
Site & Notes:		

Notes:

NGC	Filter
FOV	MAG

Notes:

NGC	Filter
FOV	MAG

Notes:

NGC	Filter
FOV	MAG

Astronomy Observation Log

Name:		Location:	Page:

Index	Details	Notes
Object	Date: / / Time : Eye Piece: Seeing: Instrument Type:	
Object	Date: / / Time : Eye Piece: Seeing: Instrument Type:	
Object	Date: / / Time : Eye Piece: Seeing: Instrument Type:	
Object	Date: / / Time : Eye Piece: Seeing: Instrument Type:	
Object	Date: / / Time : Eye Piece: Seeing: Instrument Type:	
Object	Date: / / Time : Eye Piece: Seeing: Instrument Type:	
Object	Date: / / Time : Eye Piece: Seeing: Instrument Type:	
Object	Date: / / Time : Eye Piece: Seeing: Instrument Type:	
Object	Date: / / Time : Eye Piece: Seeing: Instrument Type:	
Object	Date: / / Time : Eye Piece: Seeing: Instrument Type:	

Observation Log & Sketch Template

Observer:_____   Date:____/____/_____   Time:_____:_____

Location:_____   Seeing (1 – 10):_____ Transparency (1 – 5)_____

Object:_____   Constellation:_____

R.A_____hrs_____min Dec_____deg_____min   Magnitude:_____Size:_____

Observing Equipment :_____   Eye Pieces:_____

Filters :_____   Images:_____

Field Drawing

Low Power Ocular   High Power Ocular

Description & Notes

_____

_____

_____

_____

_____

_____

_____

Astronomy Observation Sheet

Name:	Date: / /	Time: :
Scope:	Seeing:	
Site & Notes:		

Notes:

NGC	Filter
FOV	MAG

Notes:

NGC	Filter
FOV	MAG

Notes:

NGC	Filter
FOV	MAG

Astronomy Observation Log

ame:	Location:	Page:

Index	Details	Notes
Object	Date: / / Time : Eye Piece: Seeing: Instrument Type:	
Object	Date: / / Time : Eye Piece: Seeing: Instrument Type:	
Object	Date: / / Time : Eye Piece: Seeing: Instrument Type:	
Object	Date: / / Time : Eye Piece: Seeing: Instrument Type:	
Object	Date: / / Time : Eye Piece: Seeing: Instrument Type:	
Object	Date: / / Time : Eye Piece: Seeing: Instrument Type:	
Object	Date: / / Time : Eye Piece: Seeing: Instrument Type:	
Object	Date: / / Time : Eye Piece: Seeing: Instrument Type:	
Object	Date: / / Time : Eye Piece: Seeing: Instrument Type:	
Object	Date: / / Time : Eye Piece: Seeing: Instrument Type:	

Observation Log & Sketch Template

Observer:_____ Date:____/___/_____ Time:_____:_____

Location:_____ Seeing (1 – 10): _____ Transparency (1 – 5)_____

Object:_____ Constellation:_____

R.A____hrs____min Dec____deg____min Magnitude:_____Size:_____

Observing Equipment :_____ Eye Pieces:_____

Filters :_____ Images:_____

Field Drawing

Low Power Ocular   **High Power Ocular**

Description & Notes

_____
_____
_____
_____
_____
_____
_____
_____
_____

Astronomy Observation Sheet

Name:	Date: / /	Time: :
Scope:	Seeing:	
Site & Notes:		

Notes:

NGC	Filter
FOV	MAG

Notes:

NGC	Filter
FOV	MAG

Notes:

NGC	Filter
FOV	MAG

Astronomy Observation Log

Name:		Location:	Page:
Index	**Details**	**Notes**	
Object	Date: / / Time : Eye Piece: Seeing: Instrument Type:		
Object	Date: / / Time : Eye Piece: Seeing: Instrument Type:		
Object	Date: / / Time : Eye Piece: Seeing: Instrument Type:		
Object	Date: / / Time : Eye Piece: Seeing: Instrument Type:		
Object	Date: / / Time : Eye Piece: Seeing: Instrument Type:		
Object	Date: / / Time : Eye Piece: Seeing: Instrument Type:		
Object	Date: / / Time : Eye Piece: Seeing: Instrument Type:		
Object	Date: / / Time : Eye Piece: Seeing: Instrument Type:		
Object	Date: / / Time : Eye Piece: Seeing: Instrument Type:		
Object	Date: / / Time : Eye Piece: Seeing: Instrument Type:		

Observation Log & Sketch Template

Observer:_____ Date:_____ / / _____ Time:_____:_____

Location:_____ Seeing (1 – 10): _____ Transparency (1 – 5)_____

Object:_____ Constellation:_____

R.A_____hrs_____min Dec_____deg_____min Magnitude:_____Size:_____

Observing Equipment :_____ Eye Pieces:_____

Filters :_____ Images:_____

Field Drawing

Low Power Ocular   **High Power Ocular**

Description & Notes

_____
_____
_____
_____
_____
_____
_____
_____

Astronomy Observation Sheet

Name:	Date: / /	Time: :
Scope:	Seeing:	
Site & Notes:		

Notes:

NGC	Filter
FOV	MAG

Notes:

NGC	Filter
FOV	MAG

Notes:

NGC	Filter
FOV	MAG

Astronomy Observation Log

Name:		Location:		Page:

Index	Details		Notes
Object	Date: / / Time : Eye Piece: Seeing: Instrument Type:		
Object	Date: / / Time : Eye Piece: Seeing: Instrument Type:		
Object	Date: / / Time : Eye Piece: Seeing: Instrument Type:		
Object	Date: / / Time : Eye Piece: Seeing: Instrument Type:		
Object	Date: / / Time : Eye Piece: Seeing: Instrument Type:		
Object	Date: / / Time : Eye Piece: Seeing: Instrument Type:		
Object	Date: / / Time : Eye Piece: Seeing: Instrument Type:		
Object	Date: / / Time : Eye Piece: Seeing: Instrument Type:		
Object	Date: / / Time : Eye Piece: Seeing: Instrument Type:		
Object	Date: / / Time : Eye Piece: Seeing: Instrument Type:		

Observation Log & Sketch Template

Observer:_____ Date:___/___/_____ Time:____:____

Location:_____ Seeing (1 – 10):_____ Transparency (1 – 5)_____

Object:_____ Constellation:_____

R.A_____ hrs_____min Dec_____deg_____min Magnitude:_____Size:_____

Observing Equipment :_____ Eye Pieces:_____

Filters :_____ Images:_____

Field Drawing

Low Power Ocular   **High Power Ocular**

Description & Notes

_____
_____
_____
_____
_____
_____
_____
_____

Astronomy Observation Sheet

Name:	Date: / /	Time: :
Scope:	Seeing:	
Site & Notes:		

Notes:

NGC	Filter
FOV	MAG

Notes:

NGC	Filter
FOV	MAG

Notes:

NGC	Filter
FOV	MAG

Astronomy Observation Log

ame:	Location:	Page:

Index	Details	Notes
Object	Date: / / Time : Eye Piece: Seeing: Instrument Type:	
Object	Date: / / Time : Eye Piece: Seeing: Instrument Type:	
Object	Date: / / Time : Eye Piece: Seeing: Instrument Type:	
Object	Date: / / Time : Eye Piece: Seeing: Instrument Type:	
Object	Date: / / Time : Eye Piece: Seeing: Instrument Type:	
Object	Date: / / Time : Eye Piece: Seeing: Instrument Type:	
Object	Date: / / Time : Eye Piece: Seeing: Instrument Type:	
Object	Date: / / Time : Eye Piece: Seeing: Instrument Type:	
Object	Date: / / Time : Eye Piece: Seeing: Instrument Type:	
Object	Date: / / Time : Eye Piece: Seeing: Instrument Type:	

Observation Log & Sketch Template

Observer:_____ Date:____/____/_____ Time:_____:_____

Location:_____ Seeing (1 – 10): _____ Transparency (1 – 5)_____

Object:_____ Constellation:_____

R.A_____hrs_____min Dec_____deg_____min Magnitude:_____Size:_____

Observing Equipment :_____ Eye Pieces:_____

Filters :_____ Images:_____

Field Drawing

Low Power Ocular   **High Power Ocular**

Description & Notes

_____
_____
_____
_____
_____
_____
_____
_____

Astronomy Observation Sheet

Name:		Date: / /	Time: :
Scope:		Seeing:	
Site & Notes:			

Notes:

NGC	Filter
FOV	MAG

Notes:

NGC	Filter
FOV	MAG

Notes:

NGC	Filter
FOV	MAG

Astronomy Observation Log

Name:		Location:	Page:
Index	**Details**	**Notes**	

Object	Date: / / Time :	
	Eye Piece: Seeing:	
	Instrument Type:	

Object	Date: / / Time :	
	Eye Piece: Seeing:	
	Instrument Type:	

Object	Date: / / Time :	
	Eye Piece: Seeing:	
	Instrument Type:	

Object	Date: / / Time :	
	Eye Piece: Seeing:	
	Instrument Type:	

Object	Date: / / Time :	
	Eye Piece: Seeing:	
	Instrument Type:	

Object	Date: / / Time :	
	Eye Piece: Seeing:	
	Instrument Type:	

Object	Date: / / Time :	
	Eye Piece: Seeing:	
	Instrument Type:	

Object	Date: / / Time :	
	Eye Piece: Seeing:	
	Instrument Type:	

Object	Date: / / Time :	
	Eye Piece: Seeing:	
	Instrument Type:	

Object	Date: / / Time :	
	Eye Piece: Seeing:	
	Instrument Type:	

Observation Log & Sketch Template

Observer:_____ Date:___/___/_____ Time:____:____

Location:_____ Seeing (1 – 10):_____ Transparency (1 – 5)_____

Object:_____ Constellation:_____

R.A____hrs____min Dec____deg____min **Magnitude:**_____Size:_____

Observing Equipment :_____ Eye Pieces:_____

Filters :_____ Images:_____

Field Drawing

Low Power Ocular   **High Power Ocular**

Description & Notes

_____
_____
_____
_____
_____
_____
_____

Astronomy Observation Sheet

Name:	Date: / /	Time: :
Scope:	Seeing:	
Site & Notes:		

Notes:

NGC	Filter
FOV	MAG

Notes:

NGC	Filter
FOV	MAG

Notes:

NGC	Filter
FOV	MAG

Astronomy Observation Log

Name:	Location:	Page:

Index	Details	Notes
Object	Date: / / Time : Eye Piece: Seeing: Instrument Type:	
Object	Date: / / Time : Eye Piece: Seeing: Instrument Type:	
Object	Date: / / Time : Eye Piece: Seeing: Instrument Type:	
Object	Date: / / Time : Eye Piece: Seeing: Instrument Type:	
Object	Date: / / Time : Eye Piece: Seeing: Instrument Type:	
Object	Date: / / Time : Eye Piece: Seeing: Instrument Type:	
Object	Date: / / Time : Eye Piece: Seeing: Instrument Type:	
Object	Date: / / Time : Eye Piece: Seeing: Instrument Type:	
Object	Date: / / Time : Eye Piece: Seeing: Instrument Type:	
Object	Date: / / Time : Eye Piece: Seeing: Instrument Type:	

Observation Log & Sketch Template

Observer:_____   Date:____ / ___ / _____ Time:_____:_____

Location:_____   Seeing (1 – 10):_____ Transparency (1 – 5)_____

Object:_____   Constellation:_____

R.A_____hrs_____min Dec_____deg_____min   **Magnitude:**_____**Size:**_____

Observing Equipment :_____   Eye Pieces:_____

Filters :_____   Images:_____

Field Drawing

Low Power Ocular   **High Power Ocular**

Description & Notes

_____

_____

_____

_____

_____

_____

_____

_____

Astronomy Observation Sheet

Name:	Date: / /	Time: :
Scope:	Seeing:	
Site & Notes:		

Notes:

NGC	Filter
FOV	MAG

Notes:

NGC	Filter
FOV	MAG

Notes:

NGC	Filter
FOV	MAG

Astronomy Observation Log

ame:	Location:	Page:

Index	Details	Notes
Object	Date: / / Time : Eye Piece: Seeing: Instrument Type:	
Object	Date: / / Time : Eye Piece: Seeing: Instrument Type:	
Object	Date: / / Time : Eye Piece: Seeing: Instrument Type:	
Object	Date: / / Time : Eye Piece: Seeing: Instrument Type:	
Object	Date: / / Time : Eye Piece: Seeing: Instrument Type:	
Object	Date: / / Time : Eye Piece: Seeing: Instrument Type:	
Object	Date: / / Time : Eye Piece: Seeing: Instrument Type:	
Object	Date: / / Time : Eye Piece: Seeing: Instrument Type:	
Object	Date: / / Time : Eye Piece: Seeing: Instrument Type:	
Object	Date: / / Time : Eye Piece: Seeing: Instrument Type:	

Observation Log & Sketch Template

Observer:_____ Date:____/____/_____ Time:_____:_____

Location:_____ Seeing (1 – 10): _____ Transparency (1 – 5)_____

Object:_____ Constellation:_____

R.A_____hrs_____min Dec_____deg_____min Magnitude:_____Size:_____

Observing Equipment :_____ Eye Pieces:_____

Filters :_____ Images:_____

Field Drawing

Low Power Ocular   **High Power Ocular**

Description & Notes

Astronomy Observation Sheet

Name:	Date: / /	Time: :
Scope:	Seeing:	
Site & Notes:		

Notes:

NGC	Filter
FOV	MAG

Notes:

NGC	Filter
FOV	MAG

Notes:

NGC	Filter
FOV	MAG

Astronomy Observation Log

Name:		Location:	Page:
Index	**Details**	**Notes**	
Object	Date: / / Time : Eye Piece: Seeing: Instrument Type:		
Object	Date: / / Time : Eye Piece: Seeing: Instrument Type:		
Object	Date: / / Time : Eye Piece: Seeing: Instrument Type:		
Object	Date: / / Time : Eye Piece: Seeing: Instrument Type:		
Object	Date: / / Time : Eye Piece: Seeing: Instrument Type:		
Object	Date: / / Time : Eye Piece: Seeing: Instrument Type:		
Object	Date: / / Time : Eye Piece: Seeing: Instrument Type:		
Object	Date: / / Time : Eye Piece: Seeing: Instrument Type:		
Object	Date: / / Time : Eye Piece: Seeing: Instrument Type:		
Object	Date: / / Time : Eye Piece: Seeing: Instrument Type:		

Observation Log & Sketch Template

Observer:_____ Date:____/____/_____ Time:_____:_____

Location:_____ Seeing (1 – 10):_____ Transparency (1 – 5)_____

Object:_____ Constellation:_____

R.A_____hrs_____min Dec_____deg_____min Magnitude:_____Size:_____

Observing Equipment :_____ Eye Pieces:_____

Filters :_____ Images:_____

Field Drawing

Low Power Ocular   **High Power Ocular**

Description & Notes

_____
_____
_____
_____
_____
_____
_____
_____

Astronomy Observation Sheet

Name:	Date: / /	Time: :
Scope:	Seeing:	
Site & Notes:		

Notes:

NGC	Filter
FOV	MAG

Notes:

NGC	Filter
FOV	MAG

Notes:

NGC	Filter
FOV	MAG

Astronomy Observation Log

ame:		Location:		Page:

Index	Details		Notes
Object	Date: / / Time : Eye Piece: Seeing: Instrument Type:		
Object	Date: / / Time : Eye Piece: Seeing: Instrument Type:		
Object	Date: / / Time : Eye Piece: Seeing: Instrument Type:		
Object	Date: / / Time : Eye Piece: Seeing: Instrument Type:		
Object	Date: / / Time : Eye Piece: Seeing: Instrument Type:		
Object	Date: / / Time : Eye Piece: Seeing: Instrument Type:		
Object	Date: / / Time : Eye Piece: Seeing: Instrument Type:		
Object	Date: / / Time : Eye Piece: Seeing: Instrument Type:		
Object	Date: / / Time : Eye Piece: Seeing: Instrument Type:		
Object	Date: / / Time : Eye Piece: Seeing: Instrument Type:		

Observation Log & Sketch Template

Observer:_____ Date:____ / ____ / _____ Time:_____ : _____

Location:_____ Seeing (1 – 10): _____ Transparency (1 – 5)_____

Object:_____ Constellation:_____

R.A_____hrs_____min Dec_____deg_____min **Magnitude:**_____Size:_____

Observing Equipment :_____ Eye Pieces:_____

Filters :_____ Images:_____

Field Drawing

Low Power Ocular   **High Power Ocular**

Description & Notes

_____

_____

_____

_____

_____

_____

_____

_____

Astronomy Observation Sheet

Name:	Date: / /	Time: :
Scope:	Seeing:	
Site & Notes:		

Notes:

NGC		Filter	
FOV		MAG	

Notes:

NGC		Filter	
FOV		MAG	

Notes:

NGC		Filter	
FOV		MAG	

Astronomy Observation Log

ame:		Location:		Page:

Index	Details		Notes
Object	Date: / /	Time :	
	Eye Piece:	Seeing:	
	Instrument Type:		
Object	Date: / /	Time :	
	Eye Piece:	Seeing:	
	Instrument Type:		
Object	Date: / /	Time :	
	Eye Piece:	Seeing:	
	Instrument Type:		
Object	Date: / /	Time :	
	Eye Piece:	Seeing:	
	Instrument Type:		
Object	Date: / /	Time :	
	Eye Piece:	Seeing:	
	Instrument Type:		
Object	Date: / /	Time :	
	Eye Piece:	Seeing:	
	Instrument Type:		
Object	Date: / /	Time :	
	Eye Piece:	Seeing:	
	Instrument Type:		
Object	Date: / /	Time :	
	Eye Piece:	Seeing:	
	Instrument Type:		
Object	Date: / /	Time :	
	Eye Piece:	Seeing:	
	Instrument Type:		
Object	Date: / /	Time :	
	Eye Piece:	Seeing:	
	Instrument Type:		

Observation Log & Sketch Template

Observer:_____ Date:___/___/_____ Time:____:____

Location:_____ Seeing (1 – 10): _____ Transparency (1 – 5)_____

Object:_____ Constellation:_____

R.A____hrs____min Dec____deg____min Magnitude:_____Size:_____

Observing Equipment :_____ Eye Pieces:_____

Filters :_____ Images:_____

Field Drawing

Low Power Ocular   **High Power Ocular**

Description & Notes

_____

_____

_____

_____

_____

_____

_____

_____

Astronomy Observation Sheet

Name:	Date: / /	Time: :
Scope:	Seeing:	
Site & Notes:		

Notes:

NGC		Filter	
FOV		MAG	

Notes:

NGC		Filter	
FOV		MAG	

Notes:

NGC		Filter	
FOV		MAG	

Astronomy Observation Log

Name:		Location:	Page:
Index	**Details**	**Notes**	

Object	Date: / / Time :	
	Eye Piece: Seeing:	
	Instrument Type:	

Object	Date: / / Time :	
	Eye Piece: Seeing:	
	Instrument Type:	

Object	Date: / / Time :	
	Eye Piece: Seeing:	
	Instrument Type:	

Object	Date: / / Time :	
	Eye Piece: Seeing:	
	Instrument Type:	

Object	Date: / / Time :	
	Eye Piece: Seeing:	
	Instrument Type:	

Object	Date: / / Time :	
	Eye Piece: Seeing:	
	Instrument Type:	

Object	Date: / / Time :	
	Eye Piece: Seeing:	
	Instrument Type:	

Object	Date: / / Time :	
	Eye Piece: Seeing:	
	Instrument Type:	

Object	Date: / / Time :	
	Eye Piece: Seeing:	
	Instrument Type:	

Object	Date: / / Time :	
	Eye Piece: Seeing:	
	Instrument Type:	

Observation Log & Sketch Template

Observer:_____ Date:____/____/_____ Time:_____:_____

Location:_____ Seeing (1 – 10): _____ Transparency (1 – 5)_____

Object:_____ Constellation:_____

R.A_____hrs_____min Dec_____deg_____min Magnitude:_____Size:_____

Observing Equipment :_____ Eye Pieces:_____

Filters :_____ Images:_____

Field Drawing

Low Power Ocular **High Power Ocular**

Description & Notes

_____

_____

_____

_____

_____

_____

_____

_____

Astronomy Observation Sheet

Name:	Date: / /	Time: :
Scope:	Seeing:	
Site & Notes:		

Notes:

NGC	Filter
FOV	MAG

Notes:

NGC	Filter
FOV	MAG

Notes:

NGC	Filter
FOV	MAG

Astronomy Observation Log

ame:	Location:	Page:

Index	Details	Notes
Object	Date: / / Time : Eye Piece: Seeing: Instrument Type:	
Object	Date: / / Time : Eye Piece: Seeing: Instrument Type:	
Object	Date: / / Time : Eye Piece: Seeing: Instrument Type:	
Object	Date: / / Time : Eye Piece: Seeing: Instrument Type:	
Object	Date: / / Time : Eye Piece: Seeing: Instrument Type:	
Object	Date: / / Time : Eye Piece: Seeing: Instrument Type:	
Object	Date: / / Time : Eye Piece: Seeing: Instrument Type:	
Object	Date: / / Time : Eye Piece: Seeing: Instrument Type:	
Object	Date: / / Time : Eye Piece: Seeing: Instrument Type:	
Object	Date: / / Time : Eye Piece: Seeing: Instrument Type:	

Observation Log & Sketch Template

Observer:_____ Date:____/____/_____ Time:_____:_____

Location:_____ Seeing (1 – 10):_____ Transparency (1 – 5)_____

Object:_____ Constellation:_____

R.A_____hrs_____min Dec_____deg_____min Magnitude:_____Size:_____

Observing Equipment :_____ Eye Pieces:_____

Filters :_____ Images:_____

Field Drawing

Low Power Ocular   **High Power Ocular**

Description & Notes

_____
_____
_____
_____
_____
_____
_____
_____

Astronomy Observation Sheet

Name:	Date: / /	Time: :
Scope:	Seeing:	
Site & Notes:		

Notes:

NGC	Filter
FOV	MAG

Notes:

NGC	Filter
FOV	MAG

Notes:

NGC	Filter
FOV	MAG

Astronomy Observation Log

ame:		Location:	Page:

Index	Details	Notes
Object	Date: / / Time : Eye Piece: Seeing: Instrument Type:	
Object	Date: / / Time : Eye Piece: Seeing: Instrument Type:	
Object	Date: / / Time : Eye Piece: Seeing: Instrument Type:	
Object	Date: / / Time : Eye Piece: Seeing: Instrument Type:	
Object	Date: / / Time : Eye Piece: Seeing: Instrument Type:	
Object	Date: / / Time : Eye Piece: Seeing: Instrument Type:	
Object	Date: / / Time : Eye Piece: Seeing: Instrument Type:	
Object	Date: / / Time : Eye Piece: Seeing: Instrument Type:	
Object	Date: / / Time : Eye Piece: Seeing: Instrument Type:	
Object	Date: / / Time : Eye Piece: Seeing: Instrument Type:	

Observation Log & Sketch Template

Observer:_____ Date:____/____/_____ Time:____:____

Location:_____ Seeing (1 – 10): _____ Transparency (1 – 5)_____

Object:_____ Constellation:_____

R.A_____hrs_____min **Dec**_____deg_____min **Magnitude:**_____Size:_____

Observing Equipment :_____ Eye Pieces:_____

Filters :_____ Images:_____

Field Drawing

Low Power Ocular   **High Power Ocular**

Description & Notes

_____
_____
_____
_____
_____
_____
_____
_____

Astronomy Observation Sheet

Name:	Date: / /	Time: :
Scope:	Seeing:	
Site & Notes:		

	Notes:

NGC	Filter
FOV	MAG

	Notes:

NGC	Filter
FOV	MAG

	Notes:

NGC	Filter
FOV	MAG

Astronomy Observation Log

Name:	Location:	Page:

Index	Details		Notes
Object	Date: / /	Time :	
	Eye Piece:	Seeing:	
	Instrument Type:		
Object	Date: / /	Time :	
	Eye Piece:	Seeing:	
	Instrument Type:		
Object	Date: / /	Time :	
	Eye Piece:	Seeing:	
	Instrument Type:		
Object	Date: / /	Time :	
	Eye Piece:	Seeing:	
	Instrument Type:		
Object	Date: / /	Time :	
	Eye Piece:	Seeing:	
	Instrument Type:		
Object	Date: / /	Time :	
	Eye Piece:	Seeing:	
	Instrument Type:		
Object	Date: / /	Time :	
	Eye Piece:	Seeing:	
	Instrument Type:		
Object	Date: / /	Time :	
	Eye Piece:	Seeing:	
	Instrument Type:		
Object	Date: / /	Time :	
	Eye Piece:	Seeing:	
	Instrument Type:		
Object	Date: / /	Time :	
	Eye Piece:	Seeing:	
	Instrument Type:		

Observation Log & Sketch Template

Observer:_____ Date:____/____/_____ Time:_____:_____

Location:_____ Seeing (1 – 10): _____ Transparency (1 – 5)_____

Object:_____ Constellation:_____

R.A_____hrs_____min Dec_____deg_____min Magnitude:_____Size:_____

Observing Equipment :_____ Eye Pieces:_____

Filters :_____ Images:_____

Field Drawing

Low Power Ocular   **High Power Ocular**

Description & Notes

_____
_____
_____
_____
_____
_____
_____
_____

Astronomy Observation Sheet

Name:		Date: / /	Time: :
Scope:		Seeing:	
Site & Notes:			

Notes:

NGC	Filter
FOV	MAG

Notes:

NGC	Filter
FOV	MAG

Notes:

NGC	Filter
FOV	MAG

Astronomy Observation Log

ame:	Location:	Page:

Index	Details	Notes
Object	Date: / / Time : Eye Piece: Seeing: Instrument Type:	
Object	Date: / / Time : Eye Piece: Seeing: Instrument Type:	
Object	Date: / / Time : Eye Piece: Seeing: Instrument Type:	
Object	Date: / / Time : Eye Piece: Seeing: Instrument Type:	
Object	Date: / / Time : Eye Piece: Seeing: Instrument Type:	
Object	Date: / / Time : Eye Piece: Seeing: Instrument Type:	
Object	Date: / / Time : Eye Piece: Seeing: Instrument Type:	
Object	Date: / / Time : Eye Piece: Seeing: Instrument Type:	
Object	Date: / / Time : Eye Piece: Seeing: Instrument Type:	
Object	Date: / / Time : Eye Piece: Seeing: Instrument Type:	

Observation Log & Sketch Template

Observer:_____ Date:____/____/_____ Time:____:____

Location:_____ Seeing (1 – 10): _____ Transparency (1 – 5)_____

Object:_____ Constellation:_____

R.A_____ hrs_____min Dec_____deg_____min Magnitude:_____Size:_____

Observing Equipment :_____ Eye Pieces:_____

Filters :_____ Images:_____

Field Drawing

Low Power Ocular   **High Power Ocular**

Description & Notes

_____
_____
_____
_____
_____
_____
_____
_____

Astronomy Observation Sheet

Name:	Date: / /	Time: :
Scope:	Seeing:	
Site & Notes:		

Notes:

NGC	Filter
FOV	MAG

Notes:

NGC	Filter
FOV	MAG

Notes:

NGC	Filter
FOV	MAG

Astronomy Observation Log

| ame: | | Location: | | Page: |

Index	Details		Notes
Object	Date: / / — Time :		
	Eye Piece: — Seeing:		
	Instrument Type:		
Object	Date: / / — Time :		
	Eye Piece: — Seeing:		
	Instrument Type:		
Object	Date: / / — Time :		
	Eye Piece: — Seeing:		
	Instrument Type:		
Object	Date: / / — Time :		
	Eye Piece: — Seeing:		
	Instrument Type:		
Object	Date: / / — Time :		
	Eye Piece: — Seeing:		
	Instrument Type:		
Object	Date: / / — Time :		
	Eye Piece: — Seeing:		
	Instrument Type:		
Object	Date: / / — Time :		
	Eye Piece: — Seeing:		
	Instrument Type:		
Object	Date: / / — Time :		
	Eye Piece: — Seeing:		
	Instrument Type:		
Object	Date: / / — Time :		
	Eye Piece: — Seeing:		
	Instrument Type:		
Object	Date: / / — Time :		
	Eye Piece: — Seeing:		
	Instrument Type:		

Observation Log & Sketch Template

Observer:_____ Date:____/____/____ Time:____:____

Location:_____ Seeing (1 – 10): ____ Transparency (1 – 5)____

Object:_____ Constellation:_____

R.A____hrs____min Dec____deg____min Magnitude:_____Size:_____

Observing Equipment :_____ Eye Pieces:_____

Filters :_____ Images:_____

Field Drawing

Low Power Ocular   **High Power Ocular**

Description & Notes

_____

_____

_____

_____

_____

_____

_____

_____

Astronomy Observation Sheet

Name:	Date: / /	Time: :
Scope:	Seeing:	
Site & Notes:		

Notes:	
NGC	Filter
FOV	MAG

Notes:	
NGC	Filter
FOV	MAG

Notes:	
NGC	Filter
FOV	MAG

Astronomy Observation Log

Name:	Location:	Page:

Index	Details	Notes
Object	Date: / / Time : Eye Piece: Seeing: Instrument Type:	
Object	Date: / / Time : Eye Piece: Seeing: Instrument Type:	
Object	Date: / / Time : Eye Piece: Seeing: Instrument Type:	
Object	Date: / / Time : Eye Piece: Seeing: Instrument Type:	
Object	Date: / / Time : Eye Piece: Seeing: Instrument Type:	
Object	Date: / / Time : Eye Piece: Seeing: Instrument Type:	
Object	Date: / / Time : Eye Piece: Seeing: Instrument Type:	
Object	Date: / / Time : Eye Piece: Seeing: Instrument Type:	
Object	Date: / / Time : Eye Piece: Seeing: Instrument Type:	
Object	Date: / / Time : Eye Piece: Seeing: Instrument Type:	

Observation Log & Sketch Template

Observer:_____ Date:____/____/_____ Time:____:_____

Location:_____ Seeing (1 – 10): _____ Transparency (1 – 5)_____

Object:_____ Constellation:_____

R.A_____ hrs_____min Dec_____deg_____min Magnitude:_____Size:_____

Observing Equipment :_____ Eye Pieces:_____

Filters :_____ Images:_____

Field Drawing

Low Power Ocular   **High Power Ocular**

Description & Notes

_____
_____
_____
_____
_____
_____
_____
_____

Astronomy Observation Sheet

Name:	Date: / /	Time: :
Scope:	Seeing:	
Site & Notes:		

Notes:

NGC	Filter
FOV	MAG

Notes:

NGC	Filter
FOV	MAG

Notes:

NGC	Filter
FOV	MAG

Astronomy Observation Log

		Location:		Page:
ame:				

Index	Details		Notes
Object	Date: / / Time : Eye Piece: Seeing: Instrument Type:		
Object	Date: / / Time : Eye Piece: Seeing: Instrument Type:		
Object	Date: / / Time : Eye Piece: Seeing: Instrument Type:		
Object	Date: / / Time : Eye Piece: Seeing: Instrument Type:		
Object	Date: / / Time : Eye Piece: Seeing: Instrument Type:		
Object	Date: / / Time : Eye Piece: Seeing: Instrument Type:		
Object	Date: / / Time : Eye Piece: Seeing: Instrument Type:		
Object	Date: / / Time : Eye Piece: Seeing: Instrument Type:		
Object	Date: / / Time : Eye Piece: Seeing: Instrument Type:		
Object	Date: / / Time : Eye Piece: Seeing: Instrument Type:		

Observation Log & Sketch Template

Observer:_____ Date:____/____/_____ Time:____:____

Location:_____ Seeing (1 – 10):_____ Transparency (1 – 5)_____

Object:_____ Constellation:_____

R.A_____hrs_____min Dec_____deg_____min Magnitude:_____Size:_____

Observing Equipment :_____ Eye Pieces:_____

Filters :_____ Images:_____

Field Drawing

Low Power Ocular   **High Power Ocular**

Description & Notes

_____
_____
_____
_____
_____
_____
_____
_____

Made in the USA
Coppell, TX
21 October 2024

38987097R00069